A Backcountry Journey

To our dearest friends
Rhona & Richard
Thank you for such a lovely
week.
Much love always
Chris & Bill
xxx
October 2000

A Backcountry Journey

NEW ZEALAND'S CHANGING SEASONS

TEXT AND PHOTOGRAPHY
PHILIP HOLDEN

Hodder Moa Beckett

By the same author:

Fiction

Fawn
Stag
White Patch
Razorback

Non-Fiction

Pack and Rifle
Hunter by Profession
Backblocks
The Deer Hunters
Seasons of a Hunter
The Hunting Breed
Outdoors in Australia
On Target
The Wild Pig in New Zealand
The Golden Years of Hunting in New Zealand
The Golden Years of Fishing in New Zealand
New Zealand: Hunters' Paradise
Holden on Hunting
The Deerstalkers
A Guide to Hunting in New Zealand
The Hunting Experience
Hunt South
Wild Game
More Holden on Hunting
Fall Muster
On the Routeburn Track
Along the Dingo track
In Search of the Wild Pig
Station Country I
Station Country II
Station Country III
The New Zealand Hunter

Photographs:

Front Cover: Merinos near Millers Flat, Otago.
Front Endpapers: Hereford cattle grazing Landsborough Station, Haast Pass, Westland.
Opposite Half Title: Romney hoggets about to be shorn on The Crawfords, once part of the vast Galloway run near Alexandra.
Half Title: Hereford Cattle, Landsborough Station.
Title: Midsummer, a sky of blue and a hillside alive with sweet vernal, a native grass abundant in Central Otago.
Contents Page: Looking across Upper Manorburn Dam, from Galloway Station.
Back Endpapers: Looking across the Tasman River to Mt Cook.

ISBN 1-86958-633-6

Published in 1998 by Hodder Moa Beckett Publishers Limited, [a member of the Hodder Headline Group] 4 Whetu Place, Mairangi Bay, Auckland, New Zealand
Produced, designed and typeset by Hodder Moa Beckett Publishers Ltd
Scanning and film by Microdot
Printed by Kyodo Printing Co. Ltd, Singapore

Contents

An Opening

ARGUABLY NO COUNTRY OF A COMPARABLE LANDMASS TO NEW ZEALAND OFFERS AN OUTDOORS PHOTOGRAPHER AS MUCH VARIETY AND AS MUCH SCOPE AS THIS COUNTRY. ALMOST EVERY HILL WE CLIMB, MOUNTAIN PASS WE CROSS AND COASTAL headland we look beyond opens up an entirely different panorama.

You can, for instance, travel in a very short time from a high rainfall region such as the upper Haast River in Westland to the low rainfall parts of Central Otago. In the course of the journey the landscape changes dramatically – from a big river valley dominated by heavily forested slopes and brooding mountains rearing above, to a sun-blasted, rock-studded land not unlike parts of New Mexico or Arizona, or, much closer to home, central Australia.

Wherever you journey in New Zealand, the backcountry offers breathtaking vistas. These landscapes are captured here throughout the changing seasons, offering a unique calendar of the year and a record of the people who work and live in rural New Zealand.

On Nukuhakari station, King Country, autumn.

Overleaf: Romney sheep on the Wairarapa coastline at first light, Glenburn Station.

Summer

The Summer Season

THIS WAS AN IDYLLIC SCENE IN THE BIG VALLEY OF THE NEVIS RIVER IN OTAGO, AN ALTOGETHER GORGEOUS SUMMER AFTERNOON AS HEREFORD COWS AND THEIR CALVES FANNED OUT ACROSS THE GRASSY RIVER BOTTOM. THE WINDING RIVER SPARKLED LIKE a precious blue jewel. It would eventually merge with the Kawarau River between Queenstown and Cromwell. Right then you wouldn't have wanted to be anywhere else but in the lovely valley of the Nevis River.

I adore the summer, the lovely warm-to-hot days, the drawn-out evenings, the sunrises and sunsets, oftentimes quite spectacular. I can well understand why certain tribes or peoples worshipped the sun. The sun is a tangible god.

In summer we wear considerably less clothing than at other times of the year, and perhaps because of that, we tend to feel much freer, less restricted. Again in summer we feel more optimistic, better able to cope with problems, winter's chills and winter's grey days are far from

On Morven Hills Richard Snow, who does much of the stockwork himself, musters Hereford cows and calves by helicopter.

Previous Page: Near Queenstown – below the Remarkables a field of rape forms a glorious foreground in the first month of summer.

our minds. It is as though we draw our inner strength from the beneficial rays of the sun. Is that why people who live in hot climates are more volatile and more passionate than people who do not? Either way, the summer in this country is a grand time to be eating up the miles in a four-wheel-drive vehicle.

Summertime in the backcountry on the big stations can mean many things. Certainly it has a great deal to do with the whole process of breeding stock – like, for instance, the Hereford cows and their calves in the valley of the Nevis River. They were being run close to the main homestead area of the property rather than up on the higher country where hardy wethers ranged.

Raising stock for market is an essential part of life in the country. The young animals, having been weaned, are sold off around that period when summer growth is no more and autumn is just around the corner. Indeed, the stock-carrying capacity of the land, linked with the practice of growing winter feed and the obvious possibility of having to purchase extra feed in a hard winter, dictates the sale of stock on just about any station in the country you care to name.

Meanwhile, the cattle in the Nevis Valley were further upriver, belly-deep in snow tussocks that gleamed yellow-gold in the high sun. A harrier soared above the cattle, dark against the sun; the high-pitched sounds of unseen pipits filled the air.

On other stations they were actually working with cattle as I passed through. One hot mid-summer's morning, for instance, they were dealing with 217 cross-bred, three-month-old calves on The Muller, that grand old station at the head of the Awatere Valley. Steve Satterthwaite owns this 36,500 hectare chunk of land and that is good, for he both loves and appreciates it and that isn't always the case.

At any rate, rugged-looking Steve didn't have to explain that earmarking, drenching, dehorning and castrating the bull calves was tough work in Aussie-like temperatures.

What's more, by their agitated bellowing could you fail to be aware of how unhappy the cows were to be separated from their young ones for the first time. Given time, late morning, they would

Central Otago: Stonecrop covers a low hillface near the Kawarau River, while the Cairnmuir Mountains rise in the background.

all be together again. Then later still, in mid-February, all of the station's cattle including 23 big Simmental bulls, would be taken to the head of the Acheron River (which flows south-westwards through the high country west of the inland Kaikoura Range). That meant three days in the saddle getting them there for Steve and his workforce. Great stuff! Out there, Steve explained, was real "high country", the cattle never lower than 1300 metres. Naturally all this would be under snow in winter, which explained why the cattle were brought down to a much lower level in the autumn.

The location was Fiordland National Park, late summer. I had paused in the forest to admire a southern rata, a common enough tree on the West Coast. The red flowers of the southern rata, even more deep and brilliant than the rata tree of the central North Island forests or the pohutukawa trees so plentiful in coastal regions of the more sub-tropical northlands, are indeed spectacular. But either way the summertime flowers of all of these trees bring the changing landscape wonderfully to life. Indeed, they are all so similar to each other that they surely must, in a time long past, have sprung from the same pod.

For anyone who appreciates flowering trees Central Otago would, in the summer, be particularly appealing, for this is the time when wildflowers can be seen at their best: stonecrop, red spur valerian, California poppies, meadow buttercups, and many more varieties constantly delight the traveller.

Fiordland National Park in the last month of summer – southern rata in full flower.

Matt McDonald

He was taller than most, trimly built, just twenty years old and good looking enough to have broken a few hearts along the way. His name was Matt McDonald.

"Been bulldoggin' since I was sixteen," he said matter-of-factly.

The Wanaka Rodeo was in full swing. A crowd of 3000, most of them holidaymakers, had gathered in the dusty heat to watch the day's events. Out in the ring the bullriding had started. Matt's turn would come soon enough, but in the meantime he fussed about checking his riding gear.

The rodeo circuit wasn't his fulltime job. "Hell no! Me an' Glen Wright work the Hawea mustering run for nearly eight months a year."

"Good job?"

"Reck'n."

Suddenly Matt McDonald's name came over the loudspeaker. Matt, the crowd heard, had been to the Sydney Royal Show with the New Zealand rodeo team; he had done real well and the team took second place.

"Second place, eh? Hey, that's terrific!"

Young Matt McDonald clamped his big white hat on his head.

"You reckon – hell! – there were only two teams in it, us and the Aussies!"

Above: Matt McDonald straddles a bull hell bent on unseating him. His ride lasted but a few seconds. No matter – no bones broken. There's always another bull to try your luck on.

Left: With eight national titles for saddle/bronc riding under his belt, Dave Ashford of Outram puts in a top ride at the Omarama Rodeo.

Overleaf: The annual Cromwell Merino Sale, billed as "The Gateway to Fine Wools". In the new year many stations and farms sell off their surplus and disposable stock units to generate funds to see them through the year ahead. Around twenty or so Otago stations were represented at the sale.

38

On the first day of January, calves anxiously await their appearance at the Wanaka Rodeo.

The Omarama Rodeo is held on the Andersons' Ben Omar station.

The St James Horses Sale is underway. A seven-strong Wrightson team, in blue shirts, conduct the auction of thirty-eight horses.

Red spur valerian and yellow California poppies blanket a rock-strewn hillface near Cromwell.

While Steve Satterthwaite and his Muller station staff work in the yards with cattle in the rising heat of a January midmorning, there is nothing for his fine chestnut gelding to do but relax. The same applies to the Huntaways, which know all about ranging over the Marlborough station's 36,500 hectares of Upper Awatere River country.

Looking out from the Hollyford Face across the river of the same name to the splendid Darran Mountains.

Approaching Dansey Pass from the Maniototo side – Glenshee station country. Dansey Pass cuts through the northern end of the Kakanui Mountains, linking Kyeburn in the Maniototo and Duntroon in the Waitaki valley. Although the Maori knew of this route long before the Europeans did, it bears the name of an Englishman, William Heyward Dansey, who explored it in 1855.

It was an overcast summer's day in the Mackenzie Country with mist lurking over the dry hills and the promise of long-awaited rain soon to be realised. A dull and rather colourless day – what chance a photograph on the drive between Lake Tekapo and Burkes Pass? Suddenly my wandering attention was captured by a brilliant splash of colour on the roadside – Russell lupins. Beyond the fence line, across the rocky flats, rise the mist-locked Two Thumb Range.

Previous Page: Cattle are disturbed while grazing on the lush summer growth on the banks of Ben Nevis station's sparkling Nevis River.

The sweet briar is in flower on the western side of the Lindis Pass, near the headwaters of the Lindis River. Once part of Morven Hills, this is now Breast Hill station. Like Dansey Pass, the Lindis Pass was a Maori trail.

Following Page: Meadow buttercups on the outskirts of Arrowtown.

Autumn

The Autumn Season

MID-APRIL: SOUTH WESTLAND, THE LOWER KARANGARUA/JACOBS RIVER COUNTRY. THE TIME WAS EARLY MORNING, A DISTINCT CHILL IN THE AIR, THE SUN ABOUT TO CLEAR THE INCREDIBLY STEEP MOUNTAIN FACES ON THE OTHER SIDE OF THE JACOBS RIVER.

A sudden rush of wings overhead.

"Keaaa . . . keeaaa . . ." was the cry of the only mountain parrot in the world, the wonderful, totally irrepressible kea.

Apart from the keas which soon joined us, my companions were Peter Chamberlain, a professional hunting guide and his North Island client, Lionel Berry. This was Lionel's first trip to the Southern Alps and, given the season, he really couldn't have chosen a better time.

Traditionally, the autumn, or fall as it's known in North America, is the season of the hunter. Which explains why Peter had brought Lionel here to hunt chamois, that agile mountain species introduced to this country from Austria back when such things were fashionable. Me? Well, I was there at Peter's invitation, to

Previous Page: Merinos near Millers Flat, Otago. The rolling tussocklands extend to the Lammerlaw Range.

Early morning: Keas amidst the spectacular mountain scenery of South Westland, April.

observe, but most of all to take photographs of whatever took my fancy.

In autumn the South Island mountains can be wonderful to photograph, for the mountains themselves, given the great clarity of the light, are mostly clear-cut, especially of course in the early mornings and late evenings.

Here in South Westland, however, you just never knew when a bright sunny day would change into something else entirely – the mist rolling in, wave after linking wave, from the nearby coastline, a sudden downpour of rain, even more chill in the air, the mist closing in all about you, silent, oppressive. But for now high above the headwaters of the Jacobs River the weather was good, the keas amusing, my companions intent only on the hunt. I knew very well how they felt. Around this same period, but in the Cascade River country, south of Haast, I'd listened to red stags roaring from high forested slopes. There were three of them having a good go. But today I carried a camera rather than a rifle, shooting film rather than a bullet.

The tangible results of being afield with a camera in the autumn can be seen in the following pages: keas, seemingly fearless of humans, in their Westland setting, a lone wild horse in the Kaimanawa Range at a time when public feelings about their long-term future in the central North Island were perhaps running at an all-time high, or I may have simply paused in my travels to photograph some Romney sheep on a farm in the Wairarapa.

Traditionally, the autumn is big muster time on the large South Island sheep stations – extending from the Marlborough high country, through Canterbury, and into Otago and Southland. The term used for such an event, very much a highlight of the year, is "fall muster". Essentially, fall muster means rounding up the sheep on their high summer range and bringing them down to lower country before the first real snows of an early winter – in May, say – would otherwise make the task a great deal more difficult.

Interestingly, early sheepholders in Hawke's Bay ran their sheep on the tops of both the Ruahine and Kaweka Ranges in summer and, as their more southern counterparts did, they too mustered them into one big mob and then drove them down to safer levels. Even today, it is possible in those mountains to stumble across old fenceposts and tangled fencing wire hidden in the tall tussocks which they tell a story of far different times.

The photographs in this segment taken on Mount Nicholas and Glenary relate to fall muster time, while

Out in the Inland Patea country of the central North Island – crossbred cattle ranging Ngamata station, still the largest in the North Island. The station is pretty much located between the Kaimanawa Range to the west and the Kaimanawa Range to the east.

those taken on Mount White station (they had just finished fall muster there) show wethers on their way to their winter range across the Esk River, the range of mountains seen beyond the photograph of a heading dog.

The cover photograph for this book (and which also open the autumn season) was taken on a cool but otherwise clear April morning. I was on my way from Millers Flat in Central Otago to the lower Maniototo: a dry-weather route that cuts through a broad stretch of high altitude land roughly between the Lammerlaw Range in the south and the Knobby Range more to the north-west. It's all station country out there and rather confusing, except for those who own it, to fathom out just which station you are actually on. Some of the sheep are not quite sure about that one at times, too. But it was through here that in the spring of 1858 the Shennan brothers, having taken up the lease of Galloway station, brought the first Merinos into Central Otago. And that in a nutshell was the reason I wanted to drive through it myself.

At any rate, I had put Millers Flat behind me and was climbing steadily towards a low range of hills which, in effect, contained some of the hinterland. A low hump on the skyline, according to my roadmap, added up to Mt Teviot. At just 977m, and especially by local standards, Mt Teviot was nothing to get excited about.

Suddenly, I was confronted by sheep on the road. Lots of sheep. A single man was moving them with, of course, the priceless assistance of dogs. This gent turned out to John Ferguson from Millers Flat. He was a strongly-built sort with the weathered look of a hill country musterer about him, a warm smile, and a handshake of pighunter standards. In no tearing hurry, John was more than pleased to have a yarn.

Sheep? Oh, they were mixed-aged Merinos and they had spent their summer just gone on a 2400-hectare block on the other side of Mt Teviot. John's station in fact was named Mount Teviot. Having said that, John was quick to point out than he was actually crossing Onslow View station (owned by Geoff Smith) at this point. The sheep, he went on, were being taken down to the Homestead Block at Millers Flat. All accomplished in one day.

"It's really a sort of mini-fall muster, isn't it?" I pointed out.

The man from Millers Flat, ready to get on his way with his woolly charges, smiled.

"Yeah, you could say that," he said.

The Arawata River, South Westland.

Above: These Hereford cattle, seen on the south side of the Turnbull River, belong to the Eggeling family.

Right: On his Waiatoto run, Kerry Eggeling locks a Hereford cow firmly into place. Kerry's family were among the early settlers in this area.

Station Families

They were working with Herefords at the old cattle yards on Kerry Eggeling's run in the Waiatoto River valley south of Haast. It was a typical autumn operation: the cows were being tested by Cromwell vet Richard Bishop to see if they were pregnant, and at the same time the badly distressed calves were being weaned.

Taking a break, Kerry explained that the cows would be driven up the river – a day's ride – to spend their winter on the big, grassy flats. As for the calves, they would be sold at the annual Haast cattle sale.

"Not much money in cattle these days," Kerry Eggeling admitted without rancour.

The famous Nolan family, largely based in the lovely Okuru River valley, would sell off a good number of adult cattle at the same sale. I was thinking about the Nolans when I drove down into the arresting, heavily forested valley of the Cascade River at a time when the red stags were rutting.

Prior to road transportation the Nolans mustered their cattle in the Cascade and then drove some 300 of them north to the saleyards at Whataroa. From the Cascade the cattle were taken to holding yards in the Arawhata River valley. They then pushed on to Okura. Soon they would be joined by other cattlemen, the Eggelings among them. The herd might number 800 now. By mostly following the narrow coastal strip they came to the Scotts' farm at Karangarua, near Fox Glacier. After nearly three weeks on the road they reached Whataroa.

The valley of the Okuru River – Nolan country.

The Nolans and other drovers took a two-day spell at the Scott farm at Karangarua while taking their cattle north.

In the mountains of South Westland the weather, whatever the season, can change dramatically. Later in the same day the mist chokes the valley of the Karangarua and breaks up on the main ridge that marks the edge of the watershed of the Jacobs.

Conditions are wonderfully fine in the early morning above the upper valley of the Jacobs River.

Lake Wakatipu.

Previous Spread: Seen better days – on the outskirts of the tiny settlement of Makarora. You leave the West Coast and suddenly enter Otago through a low, heavily-forested saddle in the mountains which is the Haast Pass and continue along Highway 6 to Makaroa and beyond. By the time you reach Makaroa there is already a noticeable change in the landscape, for this is the drier, eastern side of the Southern Alps.

Glencoe station country, running back to the Cardrona Range. The apple tree was most likely planted by an early settler on the Crown Terrace.

Photography is light: The wondrous diversity of autumn light is revealed in this shot, taken in the evening, of the Central Otago township of Clyde.

The Anglican Church of St Paul at Arrowtown dates to 1871.

Overleaf: Nokomai station takes in country in both Southland and Otago. This is the long, narrow Nevis Block found on the Otago side, on the eastern side of the Nevis River. Here they run halfbred sheep.

This particular Merino was reluctant to swim the North Von River and Andy is about to take it across on his horse. Simon Wright lends a hand.

Andy Sutherland of Benmore station in the Mackenzie Country, with Henry Williamson behind, on the tops of the Thompson Mountains during a fall muster on Mount Nicholas station on the western side of Lake Wakatipu. The sun has yet to break through the early morning mist filling the valley of the Mararoa River.

Heading dog, Mount White.

On Mount White in the upper Waimakariri River country – some of the station's 2000 Merinos form orderly columns en route to their winter range on the slopes of the Puketeraki Range. But first they will cross the Esk River.

Overleaf: North Rough Ridge – in April an early snowfall comes to Pennyweight Ridge station, near Becks, Central Otago.

These Romney-Perendales have just been shorn on coastal Nukuhakari station in the King Country.

They also run Angus cattle on Nukuhakari station and some of the herd can be seen grazing on a most attractive headland.

Dairy cattle, early morning, King Country.

These Romneys observed on Ngakaraka, a small property in the Wairarapa, are soon to be drenched for internal and external parasites. The Romney remains the most popular breed of sheep in New Zealand, accounting in 1997 for 58 per cent (27.5 million) of the total flock.

This bleak scene, just after daybreak, is in the valley of the Aorangi Stream which rises in the Kaimanawa Mountains and eventually links up with the Moawhango River. Red and sika deer range this country, and scattered bands of horses have run wild since the days of the New Zealand Wars.

Onenui station is located on the East Coast's Mahia Peninsula, and this is typical of its steep hill country.

Overleaf: Beyond the lone wild horse unfolds the Aorangi valley, and beyond the bluish hills can be found the largest sheep station in the North Island – Ngamatea.

Winter

The Winter Season

SINCE RETURNING TO THIS COUNTRY FROM AUSTRALIA IN 1990, I'VE SPENT MOST OF THE WINTERS IN THE MID-TO-LOWER PARTS OF THE SOUTH ISLAND. THAT MIGHT SEEM RATHER ODD FOR SOMEONE WHO IS NOT ESSENTIALLY A WINTER PERSON BECAUSE winters are cold here when compared with those experienced in the North Island. But to compensate for that there is more winter sunshine and certainly far less rain in the Southern Lakes District (where I have been mostly based) than just about anywhere north of Cook Strait.

What I like best about the winters down South is that they are clear-cut, as distinct as a razor-backed spur catching the last of the sun, as hard and as bright as a well-polished diamond. South Island winters, especially in the back country, breed a race of men with the bark on.

Having said that, it is the winters I spent as a Forest Service hunter in the 1960s that I remember best. Most of all I remember those central North Island mountain ranges – the Kawekas, the Ruahines. By mid-June, say, their lofty tops were under deep snow, the snow oftentimes extending

The lightest of June snowfalls dusts parts of the arresting landscape of the Mackenzie Country between Burkes Pass and Tekapo.

Previous Spread: Below the Torlesse Range, and near Broken River, these Merino ewes are in their winter block on Castle Hill station in Canterbury.

to the treeline, even, at times, piling up in the valley bottoms.

To stand alone in deep snow on the upper slopes or even the tops of those high places on a fine day was well worth the effort it required to get there. From certain parts of both the Kaweka and Ruahine Ranges you could, for instance, see a subdued Ruapehu; and, even further back, the cone shaped peak we then called Egmont (now, Taranaki).

Then to pause on a steep ridge in the dense Urewera country on the first clear day in a week and, through a window-like gap in the trees, gaze at the mountain known as Maungapohatu, was both satisfying and thought-provoking. Maungapohatu stood at 1336m, a square-topped mountain at the head of the Tauranga River, a part of the large watershed of the Whakatane River. Sometimes I was tempted to climb to the top and very nearly did on a day in late July when a light mantle of snow capped it. But Maungapohatu is a sacred mountain, an ancient burial place of the Tuhoe tribe, a sacred place best left well alone. A tapu had been placed on Maungapohatu. And so I never did surmount Maungapohatu and look out over what might have appeared to be everlasting forestlands.

Especially memorable in winter were the big rivers that bisected the bush-clad hills of the Urewera country. In high flood they were indeed a sobering sight. But the river I recall most in winter is the Ngaruroro, which begins its life pretty much in the Kaimanawas and ends it near Clive, a little town south of Napier. Running high, the Ngaruroro was a mud-tainted, angry giant, powerful enough to sweep the huge broken limbs of once forest kings along with the flow as though they were mere saplings that had been uprooted by whatever means and then washed down to the river itself in a time of heavy rain.

When you were camped at a hut near such a river as this in raging flood huge rocks and boulders the size of a gold miner's one-roomed cottage made an almost fearful noise as they thumped and crashed along the river bottom. It was not a pleasant sound to listen to at night! Most of us, however, still managed to get a good night's sleep.

So winter in the hills was a time of snow and rain, a time when you were very often confined

Amidst the misty, mystical mountains of the Urewera Country, home of the Tuhoe people.

to a hut for days on end: the rain hammering constantly on the corrugated iron roofing, damp clothing near a smoky fire, sometimes that cold wet bootlaces froze solid overnight. Roll on the spring, is what we used to say with much feeling.

Away from cities and towns it is, generally speaking, a rather quiet life on the big high country stations. Even so, there is always something to do: fencing close to home, maintenance work on vehicles or equipment, whatever.

But this is mostly a period when the stock, having been brought down from the high country, now reside in close proximity to the main station complex. Some stock – wethers, for instance – may well range high country.

Winter naturally means feeding out supplementary feed simply because there isn't any natural growth left. For the farmer this means a daily process, come rain, hail, or snow, that carries on until the new spring growth arrives. By long winter's end everyone is heartily sick of feeding out.

And so they were feeding out on just about all of the stations I have called in at, or simply observed from the roadside, in recent winters. I like to watch the various types of animals – sheep, cattle or deer – enjoying their daily handout. While feeding out is a real chore where the farmer is concerned it is, on the other side of the coin, the main thing to look forward to inasfar as his stock is concerned.

It was an especially rewarding trip I made to parts of Canterbury in the mid-winter of 1997, a time when real snow had yet to fall on the many mountains of the Southern Lakes District, and I'd travelled upcountry to find it.

Here in the huge valley of the Waimakariri River was another story altogether. Once I crossed Porters Pass on Highway 73 the entire landscape was under fresh snow. I pulled off the icy road a little, got out of my Trooper, and drew that clean high country air deep into my lungs.

Ahead of me the road dipped and twisted down to lower country and the snow stretched both pristine and unbroken across those semi-flatlands to the tops of the Craigieburn Range – an awe-inspiring winter landscape.

Horses free-ranging at the tiny Maori settlement of Ruatahuna, Urewera country.

Overleaf: Charolais crossbred cattle on Coal Hill station, Rangitata River.

Above: Eden Crawford and working dogs on Grasmere station. The cookshop/shearers' quarters dates to the turn of the century.

Right: Grasmere Lodge – the gabled limestone wall of the original homestead can be seen through the open gateway.

Grasmere Station

As I headed west across the Canterbury Plains towards Arthur's Pass I came to Grasmere station, nestled snugly below the Craigieburn Range. Heavy snow had fallen a few days ago and here, at the 700-metre mark, it still covered the station flatlands. There was very little movement at the station.

Manager Eden Crawford, late of Glen Wye station out near the Lewis Pass, was a tall, rangy type with an easy way about him. These days, he explained, Grasmere was just 800 hectares in size, a far cry from days long gone. Sheep numbers were down and soon there would be only cattle and deer on the station; this was significant because Merino sheep had been run there for about 140 years.

At Grasmere the old station buildings – the traditionally red-painted cookshop-come-shearers' quarters and a small cottage – were particularly interesting. They dated to about the turn of the century, Oliver Newbegin believed. He and Vicki Harraway had turned Grasmere into a high-country retreat.

Oliver pointed out that Grasmere Lodge retained the limestone walls of the original homestead. Joseph Hawdon was the first runholder, taking on the run in October 1857; the homestead was built in the following year.

Castle Hill station owner Mike Bradley feeds his ewes on a daily basis until spring growth arrives.

Looking across White Rock station with the Rangitata River on the right and, at the far end of the valley, below the mountains, Mesopotamia.

Above: Deer enjoying winter feed near Peel Forest.

Right: The Four Peaks Range makes a pleasing backdrop.

Mount Peel

Driving into the spectacular Rangitata River valley in Canterbury could, I suppose, be easily likened to entering the gates of paradise. The day was bright and sunny, no wind, a cracker frost covered many paddocks. The scene was heartland rural: sheep and cattle and deer eating - or eagerly awaiting their handout of dry rations.

Beyond Peel Forest, up the valley, I came to the first of a series of stations that once formed part of just one station: Mount Peel, taken up by John Acland and Charles Tripp in July 1855. Mount Peel - 2000 hectares of freehold and over 40,000 hectares of leasehold land - included all of the country between the Rangitata and Orari Rivers and stretched up-river as far as Forest Creek. Acland and Tripp were amongst the first men game enough to risk grazing sheep in truly mountainous country often subject to severe snowstorms. Today, the Acland family still own Mount Peel.

A little further up the valley Don Sowden was feeding out cattle: Charolais/Hereford crossbred cattle.

"Great day!" I said expansively.

"Wouldn't be anywhere else on a day like this, y'know," he said.

We were standing on Stew Point, and I had already passed through Coal Hill, according to the manager of both runs. On the runs' combined 3200 hectares they run about 300 head of cattle, along with up to 6000 halfbred sheep. Both stations are among the many runs - including White Rock and Mesopotamia - that were once part of Mount Peel station.

The Mount Peel homestead dates to 1865–66. It was built of locally fired bricks and timber felled nearby. With some modifications it remains today very much as it was when the station was at its peak.

The Church of the Holy Innocents, built for the Acland family in 1868–69, stands near the Mount Peel homestead.

Overleaf: Pruning grapevines in a light drizzle at the Villa Maria vineyard – "The Gravels" – near Fernhill, Hawke's Bay.

Typical steep hill country, bordering Tangoio forest, between Napier and Tutira.

Following a day of heavy rain, a burst of late evening sunshine brings alive a hillside overlooking Lake Tutira, Hawke's Bay.

Overleaf: Looking across Lake Ohau towards Huxley Gorge station. The snowy peaks of the Barrier Range rise above the Hopkins River valley in the centre of the photograph.

With the Ahuriri valley in the foreground, stormclouds gather over the Barrier Range.

A hard winter near Five Rivers, Southland.

Ida Valley, Central Otago: Thirsty young Herefords have to make do with a frozen waterhole.

A welcome sight following a time of drought – heavy rainclouds gather over central Hawke's Bay.

Early morning sunlight on a hard frost near Lawrence.

Feeding out, near Queenstown.

Overleaf: Across a broad band of high plain in the Mackenzie, the Gammack Range conceals the Jollie River.

Earnslaw station takes in these flats in the Dart River valley near Glenorchy. The station runs Perendale sheep, some of which are making the most of winter feed. The background is dominated by the Humboldt Mountains.

Corriedale sheep find a real feast on The Homestead station, Marlborough.

On Little Valley station, close to Alexandra, Lindon Sanders, seen here with his dog Glen and some of his top rams, runs 6300 Merinos on 10,822 hectares of low-rainfall country.

Heading dog Glen holding top Merino rams on Little Valley station, Central Otago, late winter.

Above: With the Dunstan Mountains in the background, Dan Perriam of Bendigo station near Cromwell scans the winter range of some 2000 Merinos which are about to be mustered and shorn.

Right: Snow clouds gather over the Pisa Range and Tom and his dogs can barely be seen as the mid-morning light fades rapidly.

Bendigo Station

It was the best part of a week since I'd gone along with Dan Perriam, Tom Moore and Kirsty McIntosh as they'd brought some 2000 Merino wethers down from a wintry Dunstan Range. Now it was shearing time.

Bendigo station's six-stand woolshed was modern and brightly lit. Each stand was occupied by one of Gary Bright's shearing gang. They had shorn nearly 1000 sheep so far today, including 820 stud ewes, the pick of the 6000 sheep they would handle over a period of two weeks.

The six shearers in the Bendigo shed were, like the rest of those Gary Bright employed, mostly from Otago and Southland; Vernon, however, lived "up north", at Woodville, not far from Palmerston North.

"Yeah, there's always a heap of work for me up there," Vernon, Gary's brother, said. "Thing is, I just come down here once a year for this run."

This particular gang had started their Central Otago run on the first day of August, as they had been doing since 1993, and they were due to finish in the last week of September. There were three other stations on the run: Northburn, Mount Pisa and Long Gully, all Merino properties.

Overleaf: An hour later the threat of snow seems to have abated. Tom Moore can be seen above the small figure of Dan Perriam as they combine mobs of sheep into one big bunch.

Spring

The Spring Season

ON A VIVIDLY GREEN HEADLAND ABOVE THE MUKAMUKA STREAM, AND OVERLOOKING THE SPARKLING BLUE WATERS OF PALLISER BAY, A HEREFORD COW STOOD PATIENTLY WHILE HER CALF SOUGHT WITH EAGER MOUTH ITS MOTHER'S LIFE-GIVING JUICES.

It was spring once more. New life was up and about. Yes, it was an altogether glorious time in what is the loveliest season of them all, the season of promise.

The place was Wharekauhau station, a Wairarapa coastal property owned by Bill and Annette Shaw. I've been lucky enough to have been to Wharekauhau several times and count the Shaws as good friends.

Down below the Hereford cow and her calf were narrow coastal flats, vividly green, a flush of new growth. By this route the early British runholders first took their Australian Merinos around the Coast from Wellington to the Wairarapa.

Today Romney sheep grazed on the sunny flats, the sea beyond them, the temperature in the

In October a dense carpet of green mat daisy covers much of the Tasman River valley above Lake Pukaki and stretches out towards the Ben Ohau Range.

Previous Spread: On Wharekauhau station in the Wairarapa, the sparkling blue waters of Palliser Bay provide a pleasing backdrop as a Hereford cow obliges her hungry calf.

mid-twenties. And as the Hereford cow had recently given birth to a calf, then, by the same token, so had the ewes produced an offspring.

It was pleasant to sit there in the sun and watch the lambs as they frolicked – chasing games, mostly, follow-the-leader type of stuff. Again I realised with deep satisfaction that spring was truly here. Best of all it would be many a long warm day before winter returned.

It was a miserable Saturday morning up the Awatere Valley in Marlborough. Not to worry – the misty conditions had a certain charm of their own and the mid-to-lower valley was about as rural as it gets. Maybe the weather would improve later on. The road I followed was no longer sealed. Suddenly, I came upon a tight cluster of obviously station buildings on both sides of the road. In a paddock adjacent to the road were a number of sheep. Worth a photograph?

In a light drizzle, I checked out the sheep – all rams. They eyed me with great suspicion and then, turning, bolted to the far side of their paddock. Moments later a four-wheel-drive runabout jerked to a halt alongside my transport. Two men got out.

"G'day."

I returned the greeting, and, indicating the sheep, added, "Top rams, eh?"

The older of the men smiled with no little pride at the compliment.

"Right, they are."

At any rate, this twosome was Graham Black and his son, Duncan. And this was their station – Awapiri.

"We're about to have smoko," Graham said, jerking a hand in the direction of an old homestead.

And so up the Awatere on a dismal weekend morning in spring, I enjoyed a cup of tea in the homestead and the freshly-made scones that Bev (Graham's wife) served up went down extra-well because, in a tearing hurry to get on the road early, I'd skipped breakfast.

The winter of 1997 was for once kind in the Mackenzie Country and so in October there are still reserves of hay. No matter – there is also a deer farm on the station and the hay will be put to good use. Across the valley is the Hall Range, part of Godley Peaks station.

A short time later, with the rain having eased off, we went outside to take a much closer look at the rams I'd seen earlier. They were fine-woolled sheep, the pick of their Merino flock; Awapiri is a noted stud. Some of them would soon be on display at the Christchurch Show.

In double-quick time Flick, a heading dog, rounded up the rams to his master's satisfaction and then flopped down in the grass to not only keep a sharp eye on the sheep but also to watch just what I got up to. Neat dog, Flick.

Thanking the Blacks for their fine hospitality, I carried on upvalley, the rain bucketing down, the hills lost in mist. No, it wasn't quite the weather I'd wanted.

Soon I came to Upcot. Owned by the Stevenson family since 1902, this was one of the more significant Awatere stations. The present owners were Bill and Nikki Stevenson. I found them hard at work in their big, modern woolshed. Essentially, the station's flock of wethers, having wintered over on the hill, had been mustered and brought down to the nerve-centre of the property. Here in the woolshed they were being crutched and given a general tidy-up – a process, to judge by the racket they were making, which did not meet with their approval.

So heads down and backs bent were three shearers: fulltime shearer, Geoff Boyce, Hamish Oswald (from nearby Duntroon station), and Bill's sister, Sue Popplewell.

Bill's job was to keep the pens full of sheep so that, in simple terms, there was no wasted time inasfar as the shearers were concerned. Sue was the wool handler – that is, she kept the shearing board clean, sweeping together the wool as it was removed from the sheep and then gathering it up in her arms and dumping it into an open bale. Same old story.

At one point Bill paused, nodded at a penful of worried-looking sheep.

"Rain held off long enough for us to get them all down okay." Down from the wintering block had, I think, taken them four days. They had arrived here yesterday, late, in good weather. "The rain couldn't have come at a better time," Bill added with a grin.

"Ya reckon?" was my rueful comment.

A time for daffodils. Built of local limestone, this schoolhouse at Monavale in South Canterbury dates to 1912. The tawny slopes in the background are The Hunters Hills.

Above: The Kawarau station woolshed.

Right: Near Bannockburn – a section of the Kawarau station homestead. Like the woolshed, it was built of local stone in the 1860s. This is a low rainfall region where the air is good to breathe and dampness is pretty much unknown. The lack of humidity helps to explain why both buildings are still in a sound condition so many years later.

Kawarau Station

The fourteen-roomed homestead was long and lowslung, and some of its stone walls were 50 centimetres thick. It had stood on Kawarau station since about 1860, near Bannockburn on a small stretch of flatlands lodged between two forks of Shepherds Creek. On an especially lovely day in the first month of spring, it was a most agreeable spot.

In 1895 the station stood in excess of 80,000 hectares. Called "Big" Kawarau, it took in the Carrick and Cairnmuir Ranges, some of the Old Man Range, and a substantial part of the lower Nevis River country, where it adjoined the mighty Nokomai station. From its eastern boundary at Clyde, Big Kawarau extended up the river of the same name to the Gibbston flats. The Gibbston block, nearly 70,000 hectares, was used as an out-station and the usual woolshed and other essential buildings went up there.

During the goldrush days the station supplied local miners with mutton and bread. At its peak it carried nearly 50,000 sheep, a mixed Merino/halfbred flock, and there were up to eighteen shearers on the payroll.

In 1910 the Liberal policy of forcibly subdividing the land of wealthy runholders came into effect. Big Kawarau was big no more: it was divided into sixteen separate runs.

Today Kawarau station stands at 5000 hectares. The present-day owners are John and Natalie Anderson and their only son, Richard. The station has been in the family since 1927, when John's father bought it. Both John and his son have lived on Kawarau station all of their lives.

Merino ewes await shearing at Carrick station. Legend has it that this woolshed was built using the rocks that the wagoners coming down the steep Nevis road had used as brakestones. The Carrick woolshed dates to about 1910–11, and it was the scene of much activity when I turned up. Shearing was underway and owner Don Clark, his son Jason and another shepherd were bringing a mob of two-tooth ewes into the big dusty yard adjacent to the shed.

At Upcot station, Marlborough: Sue Popplewell crutching a Merino wether.

In the valley of the Awatere River, Marlborough, looking upriver towards the Jordon Accommodation House, which dates to 1886.

On Ihungia station, north of Gisborne, musterer Will Banks attends to his mount. Horses were always a vital part of life on the historic station and a man needed to be a fair sort of horseman to get a job on the steep East Coast hills. The glory days of Ihungia station are no more – the property, as they say in those parts, "went into pines" in 1996.

Heading dog Flick, with mustering work over, takes his ease on Awapiri station.

On their Marlborough hill-country station, Awapiri, Graham Black (left) and his son Duncan are seen with the pick of their Merino flock – the prize rams.

Twin lambs on Orongorongo station, which is reached via the 'Coast Road' out of Wainuiomata.

A late afternoon for contented mothers and fat lambs on the Canterbury Plains.

Overleaf: Lake Tekapo from near the Richmond woolshed.

Huntaway Brick would much rather be on the hill hunting Merinos than here on the chain at Richmond's old woolshed.

The twelve-stand woolshed at Richmond station in the Mackenzie Country dates to around 1908. The station, on the eastern shoreline of Lake Tekapo, has been held by the Burtscher family since 1973. They run 4000 Merinos on 9500 hectares of "hard" country.

On the eastern approaches to the Lindis Pass crossbred cattle forage amongst the tussocks following a light September snowfall.

The peaks of the Hall Range can also be seen behind the old shearers' cookshop on the next station upvalley, the Fraser family's Mount Gerald. On its 21,000 hectares Mount Gerald carries 6000 Merinos and a fair-sized herd of crossbred cattle.

Last Shot

SPRING WAS OVER YET AGAIN. THE UNENDING CYCLE, SEASON TO SEASON, WAS AS INEVITABLE AS SUNRISE OR SUNSET. WHILE THE LIGHT IN NEW ZEALAND IS GENERALLY GOOD FOR PHOTOGRAPHY THROUGHOUT THE YEAR, THE AUTUMN IS arguably the pick of the seasons. This is especially true in the high country. Here the great clarity of light is ideal for taking large-scale scenic shots. It had been such a day in autumn somewhere in the lower half of the South Island. As the afternoon started to slip away with the lowering sun, I had my mind fixed on photographing red deer. Best of all would be a stag, a rutting stag, for this was the time known as the roar.

I parked my vehicle: it was time to go on foot. On the back seat was my photographic equipment. I screwed the 300-millimetre 1:4.5 lens into one of my camera bodies. Half an hour later I was moving slowly from one patch of cover to another while ahead the sun, nearly gone, was flooding a hillside, turning it to a fiery red. A light breeze fanned my face. The air was chilly.

A distant stag bellowed his ancient challenge. He was, I realised, much too far away for me to reach before darkness fell. Another stag roared. Unseen, he was close enough to make my insides flutter as though I were a Government hunter all over again.

Cupping my hands over my mouth, I made a low moaning sound.

Nothing happened for several minutes, but then the stag came into sight. Broadside to me, he looked about him, his fine rack of antlers, well-polished, etched by the wondrous, almost unreal light the tail-end of the sun was casting on the hillside beyond.

Within a few minutes the sun had gone and before that, to the sound of the camera's shutter, so too had the stag. I had my photograph, the last shot of the day.